WELCOME TO

FOB HAIKU

By Randy Brown
A.K.A. "Charlie Sherpa"

Middle West Press LLC
Johnston, Iowa

Print ISBN: 978-0-9969317-0-0
E-book ISBN: 978-0-9969317-1-7

Middle West Press LLC
P.O. Box 31099
Johnston, Iowa 50131-9428

www.middlewestpress.com

For Household-6
and the kids

CONTENTS

BASIC ISSUE

GETTING EMBED

FOB HAIKU

LESSONS LEARNED

HOME COMING

BASIC ISSUE

fragments

I am an American
fighting in the forces
which guard my country
and our way of life.

I am prepared
to give my life
in their defense.

This is my rifle.
There are many like it,
but this one is mine.

What is the spirit of the bayonet?
To kill, kill, kill
with cold blue steel.

Blood makes the grass grow green.
And the poppies appear
in Flanders Fields.

I will turn my tourniquet
to stop the flow
of the bright red blood.

I solemnly swear (or affirm)
that I will support and defend
the Constitution of the United States
against all enemies,
foreign and domestic.

I will guard everything within

the limits of my post
and quit my post
only when properly relieved.

Do not call me "sir"—
I work for a living.

So help me, God.

Grace, Ready-to-Eat

Give us this day, some shelf-stable bread,
and potable water enough to drink
and to activate the chemical heater.

And maybe a "rock or something" on which
to lean the steaming mess, entrée and all,
as depicted and described in this diagram.

Forgive us our trespasses,
for we have trespassed a lot today—
kinda goes with the territory, and the job.

And deliver us from evil,
particularly that which we have done
unto others. See also: "trespasses," above.

For thine is the kingdom,
and the power,
and the glory.

And ours is 1,200 calories of brown-bag easy living.
"Every day is a holiday. Every day is a feast."
Just go easy on the crackers. And don't eat the Charms.

Because those are bad luck.

Amen.

wait for it

Drill sergeants first teach their troops
the Zen of Hurry Up and Wait
in sundry lines outside the Post Exchange:
There, they learn to anticipate without wanting
(or wanting too much)
pogeybait, home, glory, or comfort.

War also is often more boring than not.
Then, it is scalding. Do not covet action.

Remember: Just as a watched pot never boils,
the barracks showers never truly get hot
until someone flushes.
Be on your guard.

your squad leader writes haiku

1.
Bound as fire teams:
"I'm up. He sees me. I'm down."
Action front, Leapfrog.

2.
Push through an ambush
like hornets roused out of nests
ablaze with hatred.

3.
Your weapon is jammed?!
Slap. Pull. Observe. Release. Tap.
Then Squeeze the trigger.

4.
Cover stops bullets
and concealment hides from view.
Know the difference.

5.
Take care of your feet.
Dry socks are better than sex
out here in the field.

your drill sergeant writes haiku, too

1.
Sergeant White is black!
I am Sergeant Brown! I'm white!
Do not mix us up!

2.
You are all ate up
like a soup sandwich, soldier!
Where! Is! Your! Weapon?!

3.
This smells like perf-yoom!
Drop and give me twenty, son,
if you want your mail!

4.
Now, stay there in the
front-leaning rest position
'til I get tired!

5.
There ain't no "bathroom"
within ten miles of here!
Do you mean "La Trine"?!

Jody stole your haiku tools

1.
Jody took your truck
and wrapped it around a pole
as your girlfriend sucked.

2.
A Blue Falcon flies
into your happy nest and
makes itself at home.

3.
Never give a girl
full power of attorney.
She will screw you bad.

4.
"You're the only one,"
she said. And you were, until
someone else heard same.

5.
Jody got your pay,
stole your woman and your house.
But you still have Rex.

RANDY BROWN

love sonnet to a new K-pot

You are a hard green bowl to crack apart,
inscrutable like Chinese egg-drop soup.
I trust to you my noodled self—not heart,
not groin—instead, my gray "brain-housing group."
In old steel pots, we troops could cook our grub,
or use the liners as a pail for brass.
We washed our socks and cocks in helmet tubs,
and settled on those tuffets head or ass.
Your greater weight now floats on donut foam,
and creases lines across my forehead bared—
with leathered sweatband held in place like Rome
once clipped a crown of thorns, my skull is snared.
    But, fragile shell that's spun from Kevlar thread,
    you have one purpose: Save my pounding head.

dawn patrol

there is no happiness beyond
highway speeds and distant static,
armed only with a bullet full of coffee
and a radio hungry for daytime power.

keep alert for four-legged ghosts
that graze across the dark winter fields,
while dreams of trees and barns run black
against the coming civil twilight.

Infantry blue and blaze of orange Signal:
the start of the day before the day,
before the weekend starts ...
before the first formation ...

and the call to attention.

summer convoy follows

Reflective yellow dots and dashes
stream past like Cold War fireflies
along the Dwight D. Eisenhower
National System of Interstate and Defense Highways.

At a 10-minute rest halt, we unzip our windows
to welcome in the morning air and diesel tang.
Smitty coats the windshield with anti-bug spray, while
Nelson uncouples the fuse that stops the cooling fan.

We have eight hours ahead
of heated staying-awake
and shouted conversations
through engine noise and ear plugs.

The best off-road vehicle in the world, the Humvee,
doesn't do so well at constant concrete speeds.
Anchored to our sluggish trailers, under driving sun,
we will struggle to maintain

our 100-meter intervals.

your convoy leader writes haiku

1.
Our yellow beacons
pulse like highway fireflies
taxiing toward home.

2.
Annual Training:
Once we called it "summer camp"—
now it's prep for war.

3.
Breathing diesel fumes,
the trucks cough themselves awake.
Honk! Time to mount-up.

4.
Maintain intervals.
Wear your hearing protection.
Drivers, stay awake.

5.
Bug-guts fireworks
explode across our windshields.
Ewwwww—look at that one!

laundry list

Been there,
done that:

Answered the call.
Raised my right hand.
Signed my name.
Moved out with purpose.
Pushed the ground.
Dug some holes.
Squeezed the trigger.
Learned some skills.
Paid for school.
Traveled a bit.
Wore the uniform.
Served my country.
Lived the dream.
Embraced the Suck.
Passed the test.
Did my time.

Made great friends,
lost some others.

Been there,
done that:

Got the T-shirt.

GETTING EMBED

bullet proof me

"with your shield or on it,"
Spartan women told their men.

we bought my body armor online
from a site called "bullet proof me" dot com.
cost my wife and me a couple of grand
when I left to write the war.

be sure to read for guarantees, I joked.
the carapace unpacked as pieces
sheathed in bubble wrap.

the chestnut vest is stiff with heavy plates
that might stop a few machine gun rounds,
but will also break if you drop it wrong.

mine did its job and kept me safe.
I brought it back uncracked, and good as new
to the land of no refunds and no returns.

it's not like granddad's musket—
a thing to place over one's mantel
to inspire memorable conversation.

instead, it's curled up in a box
the dry husk of an un-embedded bug.
an aegis of change—all sales were final—
a shell of my immortal self.

three cups of chai-ku

1.
I had hoped, I guess,
for something more like Starbucks,
not yellow water.

2.
We build our nations
one tea party at a time.
They serve, we protect.

3.
No one here can lead
this endless talk of action.
"Que shura, shura."

Normally a serious man

Normally a serious man,
the brigade commander gives you a hug
and later a coin.
You keep turning up like a bad penny, he says.

You have followed him
across deserts and oceans.
First in uniform, now out of it.
You dress yourself these days.

Friends downrange frequently call attention
to your color-filled wardrobe.
You are only following the rules, you tell them.
Camouflage, according to the Army, might make you a target.

The colonel's coins are numbered.
Two hundred and forty-nine have come before,
but you are a first:
once part of the tribe, but no longer in the fight.

You showed up like Justice,
who also jumped on the plane late.
He got killed while pinned down
trying to secure a helicopter crash.

You are here to share in stories like that.

The coin is worthless, of course,
but it will pay your way back across the water,
once you have found yourself
at war.

RANDY BROWN

"Why the hug?" you ask your buddies later.
It is because you are like a puppy, they say.
You remind the Old Man of better days.
You are no longer dangerous.

You are a puppy.

You are a penny.

You are home.

the embedded reporter writes haiku

1.
"Dead-weight," "straphanger"–
I carry only stories.
No guns, no camo.

2.
What's the latest score?
War is a metaphor like
a game on T.V.

3.
'Cuz he's on our side,
the fly on the green tent wall
hears only good buzz.

4.
Troop gets caught front-page
without goggles, loses face
with sergeant major.

5.
An embedded tick
needs food, safety, and access.
The media sucks.

RANDY BROWN

fighting seasons

Even a city boy from Eastern Iowa
follows the markets, like sports, on the A.M. radio
and has a vague sense of the harvests to come.

Feeder calves and pork-belly futures, forecasts for soybeans and corn
fill our diner conversations and our mouths
like bushel-bags of baseball stats, ideals and speculations.

The Cubs might finally do it this fall. And El Niño could make a
    comeback.
Into this familiar world, armed with coffee and pie,
a waitress gently probes toward our war: *Heard anything from your
    sons?*

Floods and droughts, blizzards and winds
are no strangers to the plains. We work the land,
the land works us. We do our jobs. Good weather either happens, or it
    doesn't.

It is winter in Eastern Afghanistan, but spring is coming.
There are no crops of poppies there—that's down south.
News is, the fighting will soon resume. At least, that's what the papers
    say.

Maybe this year will finally be our year.

carry on

You say you want the corpse flowers,
the stink and rot of it all,
the sand and the hamburger, the carrion seeds,
the trigger-pulls and explosive flashes that burn
innocent faces into memory.

But these are secrets and scenes I promised not to tell
you most of all
before I even left.

Forget freedom, forget terror, forget "God and Country."
Remember instead: I went to war so that you would not
eat from this particular tree.

Even the scavenging flies should realize by now
that any gaping is a trap.

Better to see and be seen,
but not heard.

Better to live and let live,
and to have let someone else

be present at the Fall.

RANDY BROWN

Mare Nostrum ("Our Sea")

Your Midwestern grandfathers fought
on North African sands
before shipping off to Italy
to crash finally upon
the mountains and beaches there:
"Algiers to Anzio."

You have even stood
in the cemetery at El Alamein,
and heard the whispers
of the toughest crust among you
remember his hallowed days
guarding similar white stones in Washington.

Now, witness people cast off
from yet another, more current war,
and history suddenly seems less about
Mediterranean rocks and hills and lives taken
than a rhythmic ebb, and flow, and
undertow ...

Let Triton try to avert his gaze.
Let all diffuse, dissolve, and
disappear in time.
Because we are not dust, but water—
moving in spaces between nations.
We are not ashes, but waves.

from a Red Bull in winter

Painter Marvin Cone wrestled with clouds
in the New Mexican desert. Camp Cody was named
after Buffalo Bill, who was also Iowa-born.

Cody died in 1917, the same year that
citizens from the Middle West were enlisted
to form the 34th Infantry "Sandstorm" Division.

Private Cone won a contest while encamped in the sand.
Inspired by expeditions to Smithsonian digs nearby,
he drew a clay water jar and the skull of a steer.

In the Second World War, Germans would be the first
to call the 34th Infantry Division the "Red Bull."
The new nickname is official now, part of our collected tale.

Professor Cone would later point to troops
on the streets of his hometown of Cedar Rapids,
taking pride in the display of his military art.

The army wears its stories on our sleeves.
Every scrap is a battle, every stitch is a past.
We are canvas, leather, dust, and blood.

At airport gates and main street parades,
the right shoulder patch carries with it
Africa and Afghanistan, Italy and Iraq.

But you are more than these threads, these fragments, those bones:
You continue the march. You are the present, armed.
You are the "Attack!"

FOB Haiku

combat patch

Another enlistment.
Another war.
Another deployment.
Another separation.
Another patch of sand.

night vision

Our Afghan brothers cannot see past the ramp
into the black wide open that is
only feet away beneath the turn of our rotors.

Our goggle eyes paint the dark
green with spinny lights and ghosts,
a Van Gogh on black velvet: "Starry Night, with guns."

We stand *Shohna ba Shohna*, shuffling off into the churning air.
We will walk down this mountain together at dawn,
stopping in each village. The Afghans will do all the talking.

We may own the night, after all,
but we are renting their country by the day.

Café Sessrúmnir

everyone knows about Valhalla:
the eternal time-share
for weary warriors gone
berserk with roidal rage
and Monster drinks.

not all appreciate, at first,
the monastic joys found elsewhere,
in this twilight hall of clockwork meals,
reflective belts,
and indoor plumbing.

but Mother Freyja claims
half the dead,
and there are many seats
here in her playground
behind the wire.

besides, why await Ragnarök
with kettlebells and grunts,
when you can work
your pickup lines
in this Fobbit coffee shop?

Welcome to FOB Haiku

1.
All that we do here
supports the Warfighter, and
Tuesday's Salsa Night.

2.
Lush desert gardens
won't promote democracy
beyond our own walls.

3.
You'd think the poo pond
would attract more mortar rounds,
but they can't hit sh--.

4.
Effin' Infantry
guys can't even spell "pogue" right.
What total Dumas!

5.
Indoor plumbing but
still "non-potable water"?
War is hell indeed.

quiet as TOC-rats

the talk in the TOC is just talk
and the sergeant major wants it quiet—
more church than circus tent.

we're not the brains of the operation;
we're more like a nervous system.
we keep things running, and people reacting.

we pass traffic by radio, Jabber, and MIRC-chat.
our burble and babble is hushed
by the air-conditioning and fluorescent buzz.

we tell stories on boards
and paint pictures for the commander.
we are Houston to his Mars.

through our bright projections, he squints
to pierce the fog of war. He will see his glories
only as shadows on our cave wall.

track the battle.
track the battle.
track the battle.

we no longer run to the sound of guns.
instead, we phone it in
and listen to reports.

our only fear is half silence:
the constant rush of static that signals
an end to our connections.

RANDY BROWN

a Radio-Telephone Operator writes haiku

1.
O.E.-2.5.4.s
spring up like dandelions.
Time to jump the TOC.

2.
Across the spectrum,
our radios skip in sync,
frequency-hopping.

3.
Listen, and you'll hear
sounds of wind and trees, *over*.
Dark rush of white noise.

4.
Use an eraser
or toothpaste to clean the crud
from points of contact.

5.
Last-calling station,
I read you Lima Charlie.
Welcome to my world.

a Forward Observer writes haiku

1.
The King of Battle!
Genuflect to indirect
fires overhead!

2.
That song about when
"caissons go marching along"?
That's Artillery.

3.
Flash against the clouds,
a rain of steel announced by
the crump of thunder.

4.
The archer's tale starts
"I have a fire mission"
and ends in an arc.

5.
Much like carpenters,
we measure twice and shoot once.
Mistakes are costly.

RANDY BROWN

love note from a drone

I had been watching you for days,
fingers hovering above the button,
waiting for release.
I am sorry I crashed your wedding.

Ancient shepherds also feared
impulses of distant gods,
who, capricious and vain, cast
thunderous judgments

in spasms of lightning—
just as a praying mantis bites
off the head of her mate
at the moment of crisis.

We will always share these points in time:
The trigger pull, the black-and-white flash,
your instant of recognition,
my bolt from the blue.

But only my driver
will remember us together—suddenly and often—
as he later speeds along
straight lines of highway

cutting through cornfields
near his Missouri home
under cloudless, lapis skies.

leaving empty

Two Buddhas once stood like giants
in the Afghan province of Bamiyan
until the Taliban blasted them down.
They had been carved into the side of a mountain
fifteen hundred years ago.

I would have liked to have seen
even the voids that remain,
but our mission changed before we arrived.
We returned the region to those who live there.
It was safe enough, I guess. Green on the map.

We moved on to Parwan and Panjshir
and I forgot about the Buddhas.
With a soldier's eyes, I have seen in other countries
churches built on temples, and mosques built on churches.
We are not Crusaders here, however. We are builders of "capacity."

My light colonel quotes a TV show:
"All this has happened before, and all this will happen again."
The West promised a surge of enlightenment,
but soon we will leave
nothing but emptiness.

Nearly fifteen years have passed
since the Taliban left.
And I have counted each day
of our boots-on-ground time,
praying for progress and last-minute peace.

I am a fool in an Army of fools,
hugging the Buddha's foot.

RANDY BROWN

dust bunnies and combat boots

The not-a-prayer rug
beside my Army bunk
guards bare soles at night.

Pictograms of Kalashnikovs,
grenades, and A.P.C.s
are part of its bazaar tapestry.

DFAC peanut butter
laces the spring I have
set under my cot, next to my boots.

The snap of the trap
summons me to my prey.
I find each of us, kneeling.

In the dark.
In a box.
Facing west.

The Sherpatudes

Not a poem, but a list of maxims by Charlie Sherpa. These are variously applicable to working in a Tactical Operations Center (TOC), newsroom, or other daycare setting:

1. Continually ask: "Who else needs to know what I know?"
2. Continually ask: "Who else knows what I need to know?"
3. Never speak with complete authority regarding that which you lack direct knowledge, observation, and/or suppressive fires.
4. Never pull rank over a radio net.
5. Let the boss decide how he/she wants to learn.
6. Let the boss decide how he/she wants to communicate.
7. "I am responsible for everything my commander's organization knows and fails to know, learns and fails to learn."
8. Know when to wake up the Old Man. Also, know how to wake him up without getting punched, shot, or fired.
9. The three most important tasks in the TOC are: Track the battle. Track the battle. Track the battle.
10. Digital trumps analog, until you run out of batteries.
11. Always have ready at least two methods of communication to any point or person on the map.
12. Rank has its privileges. It also has its limitations.
13. Let Joe surprise you.
14. Don't let Joe surprise you.
15. The first report is always wrong. Except when it isn't.
16. The problem is always at the distant end. Except when it isn't.
17. Exercise digital/tactical patience. Communications works at the speed of light. People do not.
18. Your trigger finger is your safety. Keep it away from the CAPS LOCK, reply-all, and flash-override buttons.
19. The warfighter is your customer, and the customer is always right.
20. Bullets don't kill people. Logistics kills people.
21. Knowing *how* it works is more powerful than knowing how it's

supposed to work.

22. Cite sources on demand. State opinions when asked.
23. Work *by, with,* and *through* others. It's all about empowerment.
24. Do not seek the spotlight, Ranger. Let the spotlight find you. Then, make sure to share it with others.
25. Both the Bible and *The Art of War* make this point: It's never a mistake to put oneself in someone else's boots.
26. Humor is a combat multiplier. Except when it isn't.

LESSONS LEARNED

Hamlet in Afghanistan

"To go or not to go"
    was never the question.
"To be or not to be"
    was more like "do or die" for us.
When we were all that we could be,
we did all that we could do
to find that undiscovered I.E.D.
But nothing we can ever do
will change that day in the village,
or bring back Rosie and Guido
from the dead.

So make a choice, and remember:
Over-thinking makes cowards of us all.

Kintsugi

Ashikaga Yoshimasa sent the shards away with hope
that artisans could somehow fashion
a repair for his shattered bowl.

Lacquered gold now fills its cracks;
it is stronger in the broken places.

The helmet that saved the life of Army Specialist Tom Albers
was shipped off to the procurement program executive office.
After months of analysis, it was eventually returned

to sit in a trophy case.

The MRAP, RIP'd

No one should write a sonnet to a truck
as effin' ugly as this Oshkosh box
from Afghan mud and memory unstuck—
a Trojan cow that smells of unwash'd socks.
We will park them on the banks of Lethe
and pray for rust to o'er take us all,
that once our blood and treasure runs we be
returned to our state before the Fall.
Or should we see them home with us,
a line of armored pachyderms secured
to police streets and mobs in force surplus,
now safety from all Terror be procured?
    All black in livery to there remain,
    like us, moot witnesses to Freedom's pains.

on the runways of Kabul

Each spring, the trick in tactical fashion
is to impress upon the enemy
our marvelous technology and passion
so we achieve decisive victory.
Our purses are backpacks and shoes are boots,
accessories in shows of *force majeure*,
a catalog of camouflage too cute
for insurgent defenses to endure.
And what of the pedestrian Afghan,
the man not vested in Kabul's demise?
Dress'd in *kameez*, un-uniformed, and tan,
we shall educate him in the Western guise.
    Magazines will cover what he's wearing:
    The emperor's new clothes we're sharing.

Bagram bazaar

A walk on Disney Drive results in waves
of troops too numerous to count salutes:
a full parade of green and COIN that saves
attentions for pedestrian pursuits.
In coffee shops and pizza joints bizarre,
we have a shopping mob of sorts installed—
long lines of clustered camps we trip'd too far,
America's Main Street in outlets malled.
While in safe redoubt, do not assume
that we take no pride in place or duty.
We understand our purpose to consume
all services brought by contracts' booty.
     Cached in Bagram, we have together stood,
     but war-biz wanes—so goes our neighborhood.

"Dulce et Decorum est," Redux

Spare us, this day, the clanging of metal-on-metal.
Spare us the whooping alarms, and the pumping of fists
from outward-in, bending at the elbows, swinging upward
toward our behooded skulls. Spare us the ecstasy of fumbling,
and breathless he-told-you-so's—the known-unknowns-now-
    knowns—
the blisters of smug and oozing validations. You were not there
to sniff the air with us, and we were not there to
make your world safe for democracy. Your new certitude
smells of lilacs and garlic. Your celebration
smacks of ash and yellowcake.

There are no silver linings to this cloud; the old lies
still buried, rusting, in the distant sand:
Mission accomplished.

route step, march

Soldiers are routinely ordered to break
their steps
when crossing suspensions *en masse:*
Their boots might otherwise
resonate with the natural frequency
of a structure, causing oscillations until
their uniform motions
end in whipsaw and collapse.
This science runs counter to the military mind.
Or does it?
That many, acting as one, can choose
to destroy, or to protect,
or to each momentarily walk
at his or her own rhythm
and pace
toward a common goal?

There is as much beauty in the shuffle
as there is in the march.

Bamiyan remembered

Remember those Pink Floyd laser shows,
guitars bending and flowering
at the local planetarium?

Remember the search lamp beacons
We the People shot skyward
from the dark footprints of Ground Zero?

Remember the purple cliff-face shadows
projected by the denizens of that architectural commune
during our Arizona honeymoon?

Impressions and after,
images from machines cannot hold a candle
to real time or magic past.

The Buddha never glowed like a cell phone, after all—
never even meant to be called back—but it is good to remember
that sometimes, when we gather, our conjured lights remain

like a dance by the campfire.

a sonnet, on the objective

Yours is the second hospice note this week,
friends gone from weekend wars to weakened states,
now troops at peace—no glories more to seek.
Yet do you not recall MacArthur's fate?
"Old soldiers do not die, they fade," said he.
And fade we did, when back to home released.
From weights of rucks and dust and cares were we
in missions diminished, our wives appeased.
'Cuz we were troopers once and young enough
to sleep in graves we dug ourselves in sand.
We cursed at those who thought the living rough.
We itched for actions in a foreign land.
    But now we know what rally points are best:
    surrounded by our families, at rest.

RANDY BROWN

we are the stories

we are the stories
we tell ourselves
especially
the ones we've worn out
and broken in
like boots,
for now we can march on for days
where once we would get blisters
on our soles

HOME COMING

what sacrifice has been

in airports, well-traveled souls
confuse boots with heroes
and buy us sandwiches
while flat-talking boxes buzz

with bullet-lists and mug-shots of the fallen:
3-second shrines
to soldiers they will never know
like you

this war is on us,
they want to say
thanks for your service
have a nice day

they elevate our routine dead
with casual regard and separate
us from them
with unsustaining praise

they do not grasp our names are found
on medals and on stones
and on the lips of friends who've seen
what sacrifice has been

RANDY BROWN

here and theirs

We quit our homes
to search theirs.

We left our crops
to burn theirs.

We used our guns
to take theirs.

We spent our capital
to build theirs.

And, now returned, we wonder how
we lost our freedoms

and got theirs.

Oh-Two-Early

You ask for all the meaty details
just like that Vietnamese cab driver
who dropped me off in uniform
at the Des Moines International Airport.

Oh-Dark-Thirty is way too early for questions
about whether or not I've ever shot someone.
"That's an inappropriate question," I retort,
squeezing off each syllable into the space of a dime.

I should be nicer.
I should smile in the night, and lighten my voice.
I should use this as a teaching moment,
in order to win hearts and minds at home.

No, I've never shot anyone.
But I am also not about
to tell you that.

Huginn and Muninn start a blog

we have flown all over
the world, circulating, curating,

ink-winged Thought and Desire
now recollecting here

these gaubbled gifts to you:
some beer-colored medals,

a scrapped patch of flag,
the sharp glint of toy shrapnel,

one half-hearted pendant,
our black-marble eyes.

we are tricksters by trade,
toolmakers and tellers

once young, we stood on the shoulders of gods,
whispering epaulets, epithets, rumors of war;

now smile lines and crow's-feet
are caws for celebration.

tell us what you make of them—
click and like, or subscribe.

10 haiku about a state fair

1.
Before the harvest,
at the end of summer work:
Corn-utopia.

2.
Dough drops in oil,
white sugar dusts each petal,
funnel cakes blossom.

3.
Peabody the Boar
raises his head and eyes my
porkchop-on-a-stick.

4.
Through the cattle barns,
we trod in boots and sandals,
accustomed to smells.

5.
Barkers stand Midway
between you and your money.
Sirens call, lights flash.

6.
Bumper car drivers
stand in line for more tickets,
then careen; carom.

7.
10,000 horses
burn rubber on a dirt track,
strain for a "full pull."

8.
Knife-seller cuts / deals
limited-time offers but
lifetime guarantees.

9.
Ninety-nine counties
send a fair princess each year:
Our bounty of youth.

10.
Exhibits of taste
are a matter of record
marked with blue ribbons.

drops in the funnel

In the Fun Zone, west of the sheep barn,
the zip-line carnies eject by ones and twos
those fair-goers young/crazy/bored enough to pay $10 a head
just to slide down 130 feet of rope, from tower to ground.

Meanwhile, across the street, recruiters target able-bodied passersby
for push-up fun and prizes, handing out water bottles and bumper
    stickers,
barking that Uncle Sam will toss you out of a perfectly good aircraft
absolutely free-of-charge, if you'll just step right up and raise your hand.

No felonies, please.
No neck tattoos.
No drug-users.
No drop-outs.

Say it takes 100 names and numbers for 10 good leads:
Work those 10 hard, and you might get one enlistment—
one that might still require a couple of waivers to boot.
Just hope they don't have asthma, or psych out at Basic.

Our tan trucks and tents are caked with dust,
and the wet State Fair heat feels worse than Iraq and the 'Stan rolled
    together.
Still, there are kids lining up to play in the turrets of our Humvees,
and put their butts in the gunner's sling, like we once did.

So God bless Midway America, its fried cheese and funnel cakes,
and every sweaty, dough-faced dreamer in blue jeans
who thinks that parachutist's wings and a uniform
might just be a ticket out of fly-over country.

Stand up, hook them up, and shuffle them toward the door.

RANDY BROWN

cutting strings

Like everyone else, I read the same best-selling book—
the one about the running boys and fighting kites—
before we all flew to Afghanistan for a year.
I wondered if I would see them in combat.

The Taliban had banned kites.
And dolls.
And music.
And soccer.

The Intel guys rumored something about kites
being used to signal the approach of our trucks.
It seemed a desperate and lousy choice, if true,
putting up kids and wind to do such dirty work.

Because each spool is a story, and every thread is a life
lightly anchored to the ground by a youth.
In dusty countries with ragged flags, the children seem fated
to shortly play at cutting strings and losing kites—even without the
    Taliban.

I remembered the blessed smell of dirt and spring back home,
and what it was like to fly kites with my own kids,
watching them pay out long, unbroken lines toward the sky
on the wide, open fields of Horizon Elementary.

red moon rising

The blood moon hovers a fingertip's width
above the house next door; my daughter and I brace
our camera's sullen shutter against a tree, diminishing
the shake of our hands.

There will be other times such as these,
I tell her.
You will see this again, and again,
if you know when and where to look.

Wars and presidents will come and go.
So, too, will parents and children and other first loves.
All will be eclipsed in memory, leaving you.
Remember this.

Once, my father took me out to the backyard,
to watch a rocket rise into the night, launched from Cape Canaveral.
The image in my binoculars
danced like a waterbug on fire.

Remember this.
Remember this morning.
Remember getting up too early,
and watching the moon go dark with your dad.

And I will remember the moment
you looked away from that moon,
and squealed at the stars now visible
through the threadbare autumn trees.

static

Turns out, the psychiatrist
is a former Navy Corpsman.
He says your 5-year-old problem is
that some signals can't get through.

I learned brevity on Army radios,
pushing-to-talk in 5-second bursts,
waiting a beat to hear the response,
always thinking one phrase ahead.

Instead of speaking louder, I'm told
I should dial into your distance,
quietly fine-tuning our conversations
as if I am cracking a safe.

How was your day, *over*.

Did you make any new friends, *over*.

Daddy loves you, *out*.

N.B.C.-4 report follows

Prayer flags dot our suburban terrain
like the plastic pins and Post-It notes
we once placed on acetate overlays
marking troop movements, obstacles,
and enemy actions.

Don't dig *here.*
Don't shoot *here.*
Stay off our turf.

All friendlies, be advised:
Chemical weapons have been employed
against the insurgent dandelions.

Suburbistan

The crumple and pop of small arms fire
arrives through the sliding door I have cracked
open to the spring as I breathe in
the earthy black steam of liquid breakfast.
"The sound of Freedom!" I mutter happily to the trees,
an old joke from when we first moved to Suburbistan.

The red flag will now be up on Post, and the easy morning sun will rake
long shadows behind the paper silhouettes. Sergeants will herd their
        smoking soldiers
in waves back across the road. There, they will first click in their zeroes,
then move as individuals toward their qualifying rounds.
At all times, they will take care to keep their weapons pointed up and
        downrange.
They will take all orders from the tower. Lunch will be an M.R.E.

Now routine, the weekend noise has faded into
the distant soundscape of our lives, only occasionally called to attention
like the late 9 o'clock train ("Was that the a.m. or p.m.?"),
random Friday Night football games ("Isn't it Thursday?"),
and the Civil Defense test that happens at 12 noon
the first Saturday of each month, rain or shine.

Odysseus had himself lashed to the mast,
and told his men to plug their ears with wax
to avoid the Sirens' call. Instead, I stand on my deck, listening,
sipping, wearing a bathrobe, wishing that I could grab my musket
and run toward the sound of the guns
one last time.

ACKNOWLEDGEMENTS

"Bullet proof me" first appeared in *The Pass In Review* Vol. 1, No. 1, spring 2014.

"Café Sessrúmnir" first appeared in *The Pass In Review* Vol. 1, No. 1, spring 2014.

"Carry on" first appeared in *Line of Advance* Vol. 1, No. 1, spring 2014.

"Combat patch" first appeared in *The Pass In Review* Vol. 1, No. 1, spring 2014.

"Dawn patrol" first appeared via the Veterans Writing Project's online journal *O-Dark-Thirty* May 2, 2014.

"Fighting seasons" first appeared in *Midwestern Gothic* No. 17, summer 2015.

"A Forward Observer writes haiku" first appeared via the Veterans Writing Project's online journal *O-Dark-Thirty* Oct. 19, 2015.

"Grace, Ready-to-Eat" first appeared via the online journal *Ash & Bones* April 17, 2015.

"Hamlet in Afghanistan" first appeared in *The Deadly Writers Patrol* No. 10, spring 2015.

"Jody stole your haiku tools" first appeared in *The Deadly Writers Patrol* No. 10, spring 2015.

"Kintsugi" first appeared in *So It Goes: The Literary Journal of the Kurt Vonnegut Library* No. 3, November 2014. The theme of the issue was "creative process."

"Laundry list" first appeared in *The Deadly Writers Patrol* No. 10, spring 2015.

"Leaving empty" first appeared via the online journal *Scintilla* No. 6, spring 2014. The theme of the issue was "Literature of War: At Home and Abroad."

"Love sonnet to a new K-pot" first appeared on the Veterans Writing Project's online journal *O-Dark-Thirty* Mar. 26, 2015.

"N.B.C.-4 report follows" first appeared in the anthology *Proud* *to Be: Writing by America's Warriors*, Vol. 4, Southeast Missouri State University, Cape Girardeau, Mo.

"Night vision" first appeared in the 2015 war poetry anthology *No, Achilles: War Poetry*, Water Wood Press, Huntsville, Texas.

"Normally a serious man" first appeared in *The Unofficial Anecdotal History of Challenge Coins*, Tayler Corp., Orem, Utah, April 2015.

"Quiet as TOC-rats" first appeared in *The Pass In Review* Vol. 1, No. 1, spring 2014.

"A Radio-Telephone Operator writes haiku" first appeared in the 2015 anthology *Proud to Be: Writing by America's Warriors*, Vol. 4, Southeast Missouri State University, Cape Girardeau, Mo. It received a honorable mention in poetry.

"Red moon rising" first appeared on The Good Men Project website Sept. 28, 2015.

"The Sherpatudes" first appeared on the *Red Bull Rising* blog on Mar. 2, 2012.

"Suburbistan" first appeared in the 2014 anthology *Proud to Be: Writing by America's Warriors*, Vol. 3, Southeast Missouri State University Press, Cape Girardeau, Mo.

"10 haiku about a state fair" first appeared in the *Corn Belt Almanac*, The Heart & The Hand Press, Philadelphia, June 2015.

"Wait for it" first appeared in *The Pass In Review* Vol. 1, No. 2, summer 2014.

"We are the stories" first appeared in *Spillway* No. 23, June/July 2015. The theme of the issue was "everyday epiphanies."

"What sacrifice has been" first appeared in the 2012 anthology *Proud to Be: Writing by America's Warriors*, Vol. 1, Southeast Missouri State University, Cape Girardeau, Mo.

"Your convoy leader writes haiku" first appeared in *Line of Advance* Vol. 1, No. 1, spring 2014.

"Your drill sergeant writes haiku, too" first appeared in *The Deadly Writers Patrol* No. 10, spring 2015.

"Your squad leader writes haiku" first appeared in *The Pass In Review* Vol. 1, No. 2, summer 2014.

NOTES

REGARDING STYLE & VOCABULARY

In this collection, I hope to communicate something of the times and places I experienced while in the military. That effort starts with the vocabulary. Language in the military is a stew of ever-changing labels, buzzwords, nomenclatures, phrases, and slang. The terms and jargon I use are that of someone who joined the U.S. Army in the late 1980s, and served until late 2010. To illustrate the changes I witnessed during this time: As a recruit, I started out wearing olive-green utility fatigues, a World War II-style "steel pot" helmet with woodland camouflage cover, and polished black boots. I ended my career wearing gray-green digital camouflage, a bulletproof Kevlar helmet, and no-polish tan boots.

The military changes words as often as it changes uniforms. Often—but not always—this is driven by changes in technology, strategy, or official messaging. I joke that "every time an acronym changes, a full-bird colonel gets his wings." Someone up high arbitrarily decides that "Load-Carrying Equipment" (L.C.E.) is now called "Load-Bearing Equipment" (L.B.E.). The O.E.-2.5.4. antenna replaces the O.E.-2-Niner-2. In a briefing, the Old Man publicly squashes the use of the term "occupation" to describe our mission in Iraq. "All this has happened before, and all this will happen again."

A particular poetic challenge is to communicate how military acronyms and initialisms are to be read aloud. Abbreviations are constructed with the first letters of words and are pronounced as a word themselves. (Examples: RADAR and LASER.) Initialisms are constructed in the same way, but each letter is pronounced independently. (Examples: A.T.V. and S.U.V.) In both poems and notes, to assist readers in deciphering how such terms are to be said or read, I have added punctuation to indicate initialisms. This is not, of course, how these words might be presented elsewhere.

Where potentially helpful, I have also italicized non-English words and phrases.

In all of this, my objective is to clearly communicate across military branches, experiences, generations, and the civil-military divide. I hope that the techniques described and used here will bridge potential gaps in understanding, and make these stories accessible to new audiences.

BASIC ISSUE

"Grace, Ready-to-Eat": A "Meal, Ready-to-Eat" (M.R.E.) is a high-calorie field ration developed and used by the U.S. military.

"Wait for it": The term "pogeybait" is military slang for contraband candy, snacks, or other non-issued food, often smuggled into the field or barracks.

"Your squad leader writes haiku": Soldiers are taught to perform immediate action on their rifles to clear a jammed round, using the acronym SPORTS. The mnemonic stands for:

Slap [the magazine].
Pull [the charging handle].
Observe [the ejected round and chamber].
Release [the charging handle].
Tap [the forward-assist].
Squeeze [the trigger].

"Jody stole your haiku tools": In military marching cadences and other U.S. Army oral traditions, the fictional character Jody is described as a civilian freeloader who will steal a soldier's belongings and romantic interests while one is away from home. Marching cadences are themselves called "Jodies." The term "Blue Falcon" is used to describe someone who routinely seeks to get ahead at the expense of his or her fellow soldiers—a "B.F." In other words, a "Buddy F--ker."

"Love sonnet to a new K-pot": In the 1980s, the U.S. Army replaced the World War II-era M1 "steel pot" helmet with the heavier PASGT (pronounced "pass-GET") helmet. The acronym stands for "Personnel Armor System for Ground Troops." In reference to the PASGT's Kevlar construction, soldiers sometimes referred to the helmets as "K-pots."

GETTING EMBED

"Three cups of chai-ku": An Afghan *shura* (pronounced "SHUR-ah") is a public consultation among leaders, often conducted at a community or village level. The phrase *"que shura, shura"* is intended as wordplay on the faux Spanish/Italian phrase, *"que sera, sera"* ("Whatever will be, will be"), once popularized by the American singer Doris Day.

"Normally, a serious man": Staff Sgt. James A. Justice, 32, of Grimes, Iowa, a member of Alpha Troop, 1st Squadron, 113th Cavalry Regiment (1-113th Cav.), was killed April 23, 2011 by small arms fire in Kapisa Province, during an helicopter-borne mission to secure the position of a downed U.S. attack helicopter. He had arrived in Afghanistan as a volunteer replacement in February 2011, joining his fellow Iowa Army National Guard soldiers, who deployed in October 2010. From July to September 2011, I worked as part of the same stateside team as Justice, helping prepare more than 3,000 Iowa Army National Guard soldiers of 2nd Brigade Combat Team (B.C.T.), 34th Infantry "Red Bull" Division for deployment to Afghanistan. I knew him to be devoted and tireless in preparing training and equipment for the deploying troops. His friends called him "Juice."

"Mare Nostrum ('Our Sea')": On Feb. 11, 2015 United Nations officials announced that 300 migrants died in an attempt to cross the Mediterranean Sea, from Libya to Italy. International Organization for Migration officials estimated that, in 2014, more than 3,000 migrants died in similar attempts. Media attention to the situation was sporadic, despite the movement of approximately 170,000 people toward Europe in 2014. In October 2014, Italy suspended a search-and-rescue operation called *Mare Nostrum*, which actively sought out imperiled ships along the Libyan coast. It was replaced by a less-resourced mission fulfilled by the European Union, called Triton, which monitored areas closer to European shores.

"From a Red Bull in winter": The official motto of the U.S. 34th Infantry "Red Bull" Division is "Attack! Attack! Attack!"

FOB HAIKU

"Combat patch": In the U.S. Army, soldiers wear a patch identifying a current unit of assignment on the *left* shoulder of their Army Combat Uniforms (A.C.U.)—the camouflage field uniform. This unit patch may change as a soldier transfers to different organizations. Soldiers who have deployed to a war zone are permanently authorized to wear their deployed unit's (or headquarters') distinctive shoulder patch insignia on the *right* sleeves of their combat uniforms. This authorization continues through their military careers, and the patches do not change. By "reading" the Army uniform—left sleeves and rights sleeves—one can tell whether a soldier has gone to war, and with whom.

    "Night vision": Lacking the night-vision gear of their U.S. counterparts, Afghan soldiers on 2011's Operation Bull Whip reportedly could not see into the inky black beyond the ramps of their transport helicopters. The Dari phrase *"Shohna ba Shohna"* means "Shoulder to Shoulder," and was the motto for the NATO Training Mission-Afghanistan (N.T.M.-A.) in 2011.

    "Café Sessrúmnir": In Norse mythology, warriors killed in battle are received either in the heaven-like hall of Valhalla by the god Odin, or in the hall of Sessrúmnir (which translates as "seat-room") by the goddess Freyja. A "Fobbit" (a term that rhymes with the comfort-loving "Hobbit" race of fantasy characters created by J.R.R. Tolkien) is a soldier assigned to a support role, such as logistics or administration, who lives and works in the relative safety of a Forward Operating Base, or "FOB." Larger FOBs boast pizza parlors and coffee shops.

    "Welcome to FOB Haiku": A FOB is a Forward Operating Base, a deployed but relatively safe environment. A pogue (pronounced "pohg") is a support soldier, one not directly engaged in fighting an enemy force. In the slang of earlier conflicts, this term might be synonymous with REMF (pronounced "REM-fff")—"Rear Echelon Mother F--ker." In current conflicts, the term is nearly synonymous with "Fobbit." Recent generations of soldiers have taken to retroactively treating "pogue" as an acronym: POG (still pronounced "pohg"), or "Person Other than Grunt."

"Quiet as TOC-rats": A Tactical Operations Center (TOC, pronounced "tahk") is an organizational nerve center, responsible for monitoring and controlling current operations of a unit. This collection of people, computers, and communications equipment is similar to a "mission control" room made familiar to the public by the U.S. space program. Both Jabber and MIRC-chat are instant-messaging computer applications. The acronym MIRC stands for "Microsoft Internet Relay Chat."

"A Radio-Telephone Operator writes haiku": An O.E.-2.5.4. is a multi-directional radio antenna with a 42-foot-high sectional mast. A TOC is a Tactical Operations Center—a unit's command nexus. The phrase "Lima Charlie" is radio jargon for "loud and clear."

"Leaving Empty": The phrase "hugging the Buddha's foot" is a colloquialism for last-minute prayer.

"Dust bunnies and combat boots": A Kalashnikov is any one of a series of rifles, such as the A.K.-47, based on an original design by Russian arms inventor Mikhail Kalashnikov. The intialism A.P.C. stands for "Armored Personnel Carrier." In the U.S. Army, the acronym DFAC, pronounced "DEE-fack," stands for "dining facility"—what earlier generations of soldiers called a "chow hall."

LESSONS LEARNED

"Hamlet in Afghanistan": The initialism I.E.D. stands for "Improvised Explosive Device." This is an explosive mine, constructed using parts such as appliance timers, cellular telephones, fertilizer, and/or repurposed artillery shells.

"Kintsugi": Kintsugi is the Japanese art of repairing ceramics with lacquered resin mixed with metallic powders. Originating in the 15th century, the practice celebrates an object's history and imperfections, while also keeping it in service.

"The MRAP, RIP'd": Pronounced "EM-rap," the adjectival acronym MRAP stands for "Mine-Resistant, Ambush-Protected," and describes a modern class of armored ground vehicle that is designed to withstand blasts from Improvised Explosive Devices (I.E.D.). Pronounced "rip," the military acronym RIP stands for "Relief in Place," a term describing a period during which one deployed unit replaces another in a tactical situation.

"On the runways of Kabul": A "show of force" is a military demonstration intended to deter enemy action, such as a low-and-loud flyover by a jet; *force majeure* is a legal term potentially synonymous with an "act of God" or other overpowering force. A *kameez* is a long shirt or tunic, split at the sides below the waist and worn with trousers *(shalwar)*, and is traditional dress in Afghanistan, Pakistan, and other countries.

"Bagram bazaar": Pronounced "koyn," the acronym COIN is military jargon for "counter-insurgency." Disney Drive, a major paved thoroughfare on Bagram Airfield, is named after U.S. Army Spc. Jason A. Disney, 21, of Fallon, Nev., killed in Afghanistan Feb. 13, 2002.

"'Dulce et Decorum est,' Redux": The title evokes William Owen's "Dulce et Decorum est," a poem published posthumously in 1920, which describes a gas attack during World War I. The Latin phrase *"Dulce et decorum est pro patria mori"* is taken from the Roman poet Horace, and translates as "It is sweet and fitting to die for one's country." In October 2014, an investigation by the *New York Times'* C.J. Chivers revealed that the U.S. military had covered up discoveries by U.S. troops of chemical munitions during Operation Iraqi Freedom. (Soldiers were even injured by their encounters with rusting artillery shells containing nerve and blister agents. Their injuries often went undocumented.) Some argued the newly acknowledged weapons retroactively vindicated the U.S. invasion of Iraq, a view that chooses to forget the shifting rhetorical sands upon which our involvement there was justified.

"Bamiyan remembered": Working with approvals from UNESCO and Afghan authorities, a Chinese couple in June 2015 used 3-D laser-light projections to temporarily "restore" the presence of two giant 6th century Buddha statues, which were once carved into the mountainside in Afghanistan's Bamiyan Province, and were destroyed by the Taliban in 2001.

"A sonnet, on the objective": In January 2014, two former Iowa Army National Guard colleagues of mine independently succumbed to illnesses. Their deaths were not related to their respective military service. Lino Gerardo Bernal, 46, of Des Moines, Iowa, died Sat. Jan. 18, 2014. Roger J. Puetz, 56, of Anamosa, died Tues., Jan. 21, 2014. Each was a memorable and respected leader, peer, and citizen-soldier.

HOME COMING

"Oh-Two-Early": The phrase "Oh-Dark-Thirty" is military slang for any specified time before sunrise.

"Huginn and Muninn start a blog": In Norse mythology, the ravens Huginn (pronounced "HYOO-jin," and translated as "Thought") and Muninn (pronounced "MYOO-nin," and variously translated either as "Memory" or "Desire") daily patrol the earthly realm, then land on Odin's shoulders at dinnertime to report their findings.

"N.B.C.-4 report follows": A "Nuclear, Biological, and Chemical Report 4" is a 12-line standardized U.S. Army message format for describing a contaminated area over a radio or telephone.

THANKS

Thanks to my wife and kids. In military parlance, you are always my "main effort."

Thanks to Travis L. Martin and David P. Ervin, respective founding and current presidents of the non-profit Military Experience & the Arts, as well as the cohort of fellow travelers I have met through that organization. Thanks particularly to members Jason Poudrier, Suzanne Rancourt, and Tara Leigh Tappert for their regular insights and online camaraderie.

Thanks to founder Ron Capps, *O-Dark-Thirty* Managing Editor Jerri Bell, and others at Veterans Writing Project, for their efforts on behalf of writer-veterans everywhere.

Thanks to James Burns, one of the founders of the 2012 Sangria Summit military writers' conference, for his past financial and moral support of the *Red Bull Rising* blog.

Thanks to the mysterious DoctrineMan!!, for daily validating that "humor is a combat multiplier …" and that snark can be a force for good. At least on Facebook.

Thanks to Jeffrey Courter (with Kathleen Kruse), Kanani Fong, Lynnis Honaker, Mari Paxford, Benjamin Tupper, and other mil-bloggers, who each reached out to support a deploying citizen-soldier and his blog.

Thanks to David Stanford at Doonesbury's *The Sandbox*, a now-archived online digest of modern military writing, for his support and mentorship of the *Red Bull Rising* blog.

Thanks to Tom Ricks of the *Best Defense* blog at *Foreign Policy* magazine, for his continual support of the *Red Bull Rising* blog.

Thanks to Jeanetta Calhoun Mish, director of the low-residency Red Earth MFA in Creative Writing program at Oklahoma City University, for her ongoing support of writer-veterans, as well as the program's past financial support of the *Red Bull Rising* blog.

Thanks to Steven Wingate, coordinator of South Dakota State University's annual Great Plains Writers' Conference in Brookings, S.D., and the many writing friends first met there. Conversations started at 2014's "Coming Home: War, Healing, and American Culture" event continue to resonate daily in my life, work, and social media. The Midwest is indeed a small town.

Thanks to Susan Swartout and her colleagues at Southeast Missouri State University Press, who, through a partnership with the Missouri Humanities Council, have published the annual anthology *Proud to Be: Writing by American Warriors* since 2012.

Thanks to Emma Johnson and John Mikelson of the non-profit Writing My Way Home, providers of weekend writing workshops for veterans, particularly for the 2011 prompt that led me back to poetry.

Thanks to *Des Moines* (Iowa) *Register* reporter Tony Leys and photographer Rodney White, who were first to get eyes and boots on the ground with Iowa's 2nd Brigade Combat Team, 34th Infantry "Red Bull" Division in Afghanistan, and who also lent me their phone.

Thanks to Joy Riggs, for her friendship and editorship.

Thanks to all editors of literary journals who open their pages to explore the relationships among our civil society and the military we employ, whether in fiction, non-fiction, essay, interview, or poetry.

Finally, thanks to the men and women of the 34th Infantry "Red Bull" Division, past and present, and to their families. You are the *"Attack! Attack! Attack!"*

ABOUT THE WRITER

In 2010, Randy Brown was preparing for deployment to Eastern Afghanistan as a member of the Iowa Army National Guard's 2nd Brigade Combat Team (B.C.T.), 34th Infantry "Red Bull" Division. Since its organization in 1917, the division has historically comprised citizen-soldiers from Minnesota, Iowa, North and South Dakota, and other Midwestern states. In news reports, the 2010 deployment of more than 3,000 was billed as the largest activation of Iowa troops since World War II.

After a paperwork SNAFU dropped Brown from the list, he retired with 20 years of military service and a previous overseas deployment. He then went to Afghanistan anyway, embedding with Iowa's Red Bull units as a civilian journalist in May-June 2011. A former editor of community and metro newspapers, as well as national trade and consumer magazines, he is now a freelance writer based in Central Iowa.

He writes about military topics at: www.redbullrising.com.

Brown was the 2015 winner of the inaugural Madigan Award for humorous military-themed writing, presented by Negative Capability Press, Mobile, Ala. He was the 2012 winner of the Military Reporters and Editors' (M.R.E.) independent-blogging category, and was a finalist in the Milblogging.com awards' veteran (2011) and reporter (2012) categories.

He is the current poetry editor at the literary journal *As You Were*, published twice a year by the non-profit Military Experience & the Arts. He is also a member of Military Reporters & Editors, the Military Writers Guild, and the Military Writers Society of America.

ABOUT THE ARTIST

Cover artist Aaron Provost is an Iraq War veteran and Purple Heart recipient. At the age of 18, in July 2001, he enlisted in the U.S. Army. He served in Iraq 2004-2005 as a fire support specialist with 1st Brigade, 1st Cavalry Division. Upon leaving the service in 2007, he joined the fire department in Killeen, Texas, before eventually moving to Maryland, where he pursued his degree.

"Pushing aside many self-doubts, I set my heart to what had always been so important to me," he writes. "Art had been a part of me since as early as I can remember. After going through so much, it was the thing I had been doing all along. I was able to find myself again. From second-grade spaceship drawings, to teen angst filled sketches during high school, to terrain sketches on the observation post, to sketchbooks filled with images of my dogs ... it took me so many uncertain years to figure it all out."

Provost is a 2013 graduate of the Maryland Institute College of Art, Baltimore. The pencil sketch of a Mine-Resistant Ambush-Protected (MRAP) truck, featured on the cover of this book, comes from his senior thesis.

For more of his work, visit: www.aaronprovost.com.

Provost currently works as a graphic designer and dismount at Ranger Up, based in Durham, N.C. There, he is involved in the production of patriotic T-shirts, military-themed gear, and zombie movie chaos. Visit: www.rangerup.com.

www.ingramcontent.com/pod-product-compliance
Lightning Source LLC
Chambersburg PA
CBHW031631040426
42452CB00007B/771